THE FIRST AMENDMENT IN THE INFORMATION AGE:
Regulation & the Videotex-Teletext Industry

BY
M. JOEL BOLSTEIN

Co-published by
FREEDOM OF EXPRESSION FOUNDATION
THE MEDIA INSTITUTE
Washington, D.C.

The First Amendment in the Information Age:
Regulation & the Videotex-Teletext Industry
Copyright 1987, Freedom of Expression Foundation
All Rights Reserved.
First printing September 1987.
Co-published by the Freedom of Expression Foundation, Washington, D.C., and The Media Institute, Washington, D.C.
Printed in the United States of America.
ISBN: 0-937790-36-2

Table of Contents

Acknowledgments   v

Introduction   1

I.   Current State of Videotex and Teletext Service Industries   3

II.   Rationales for Regulating Electronic Publication   7

III.   History of Radio and Television Regulation   9

IV.   Broadcast Regulations Based on Scarcity of Frequencies   13

V.   Broadcast Regulations Based on Pervasiveness   21

VI.   Regulation of Cable Television   23

VII.   Regulation of Common Carriers   27

VIII.   Application of Broadcast Content Controls to Teletext   31

IX.	Application of Broadcast Content Controls to Videotex: An Area of Great Uncertainty	39
X.	Impact of Content Controls	45
XI.	Conclusions	49
XII.	Recommendations	51
End Notes		53

Acknowledgments

The Freedom of Expression Foundation wishes to express its thanks to CNR Partners for their seed grant that helped make this study possible.

The Media Institute and the Freedom of Expression Foundation appreciate the comments by Mr. George Shapiro of Arent, Fox, Kintner, Plotkin & Kahn, who reviewed an early draft of the manuscript.

Additionally, we wish to thank Bonneville International for their generous contribution toward the publication of this work.

Introduction

This report by the Education and Research Fund of the Freedom of Expression Foundation presents findings and conclusions based on a three-phase study of the applicability of government content regulations to teletext and videotex services. This project is part of the Foundation's comprehensive investigation of the impact of governmental regulation of the media upon American society, upon its institutions, and upon select groups within that society.

Teletext and videotex have tremendous market potential for broadcasters, equipment manufacturers, retailers, advertisers, and communications entrepreneurs. Development of these technologies would also greatly benefit society in general by expanding the marketplace of ideas, and benefit consumers in particular by providing information and convenient services.

The report begins with a description of the current state of the teletext and videotex industries. It then examines the rationales for regulating the content of broadcast, cable, and telephone transmissions. Moreover, it analyzes the government content controls applicable to those communications media, assesses their possible application or extension to teletext and videotex, and examines the possible impact of such an action on the development of these technologies. Finally, the report offers conclusions and makes recommendations for the future of teletext and videotex.

The Current State of the Videotex and Teletext Service Industries

Videotex Defined
Videotex is a two-way interactive computer-based information service in which the user is linked to a database or electronic bulletin board by telephone line or two-way cable. Most services operating in the United States rely primarily on personal computers with modem attachments to access videotex services via telephone lines. Videotex allows businesses and consumers to conduct transactions, send and receive electronic messages, and gain access to a wide variety of information. Videotex uses include electronic banking, home shopping, travel scheduling, and communication of up-to-the-minute news, sports, and financial information.

The Current State of Videotex
According to the Videotex Industry Association (VIA), there are approximately 700,000 home subscribers accessing 40 fee-based consumer videotex systems in the United States, with another 500,000 users of free public access bulletin board systems.[1] While current estimates are that videotex will grow 40 percent in 1987, 1986 was a difficult year for the industry. Many major newspaper publishers withdrew from the videotex market. Knight-Ridder shut down its Viewtron operation

after spending over $50 million during the seven years it developed and marketed the service in Florida and, later, nationwide. Times Mirror also abandoned its Los Angeles-based Gateway videotex project at a loss of almost $30 million.[2] Neither company was successful in developing a substantial customer base—Knight-Ridder had nearly 20,000 customers while Times Mirror had about 5,000.

Despite these setbacks, videotex continues to grow at a modest pace.[3] CompuServe and The Source continue to be the most viable and commercially successful videotex services. Recently, two new videotex services entered the national market, one owned by General Electric Information Services and the other by Quantum Computer Services. These services have acquired 75,000 subscribers after one year in business.[4]

Telephone companies have shown an interest in entering the videotex services market. Pacific Bell recently tested a California service called Project Victoria, which provided electronic mail, local information from the San Francisco *Chronicle*, home banking, and educational data. AT&T is experimenting with videotex technology in partnership with Chemical Bank, Bank of America, and Time, Inc. Their project, called Covidea, provides home banking and brokerage services, and news and information from the *New York Times*, *Money* magazine, and other publications. Greater involvement of the phone companies, however, is dependent on further deregulation of the telecommunications industry and the disposition of recommendations to permit the regional Bell operating companies (BOCs) to enter the information marketplace. The United States Department of Justice's recent filing with the U.S. District Court reviewing the Modified Final Judgment in the AT&T divestiture proceeding made three recommendations. First, to allow the BOCs to provide interexchange services outside their given regions. Second, to permit BOC manufacturing of equipment. And finally, to allow the BOCs to enter the information services market, which would include electronic publishing and database services.

IBM and Sears are partners in Trintex, which is planning to offer a diverse blend of videotex services. Additionally, Citicorp, NYNEX, and RCA have formed CNR Partners and are experimenting with videotex. Many other companies are investing substantial sums in development and marketing of videotex services.[5]

Outside the United States, videotex is being used in approximately 30 countries.[6] France has been the most successful at developing and marketing videotex services. It is estimated that 4.5 million French men and women use Minitel, the state-run videotex network.[7] The

Minitel service relies heavily on dedicated videotex terminals to access information.

Teletext Defined

Teletext is a one-way electronic publishing service in which textual and graphic material is delivered to the home as part of the television broadcast or over cable. Teletext uses the vertical blanking interval of the standard television signal to transmit pages of text which are formatted on the user's television screen by a decoding device.

It is possible to offer a number of services using teletext. Programs that are "closed captioned" for the hearing impaired use the vertical blanking interval. Teletext also provides standard information, such as news, weather, financial, and entertainment.

Teletext information can be continuously updated because it is computerized information that is constantly broadcast. By using the decoder, consumers can select the exact page of text they would like to view. Unlike videotex, which can handle thousands of pages of information, most teletext systems can only handle 100 or 200 pages. The advantage of teletext, however, is that it is a relatively inexpensive method of transmitting information and data from a central point to many locations.

Current State of Teletext

Teletext is not broadly available in the United States. The lack of affordable decoders and the presence of incompatible teletext standards are the main reasons teletext is not reaching many American consumers.

There are, however, numerous ongoing teletext ventures. Taft Broadcasting's WKRC-TV in Cincinnati, Ohio produces a teletext service and distributes it nationally over the vertical blanking interval of superstation WTBS in Atlanta.[8] Jefferson-Pilot station WBTV in Charlotte, North Carolina produces its own local teletext magazine.[9]

NBC launched its national teletext service on May 16, 1983. The service consists of an 80-page magazine with sections devoted to world and national news, sports, weather, life-style, financial news, horoscopes, games, and a travel catalogue.[10] KTTV in Los Angeles started its teletext operation during the 1984 Summer Olympics, giving Olympic results and traffic information. After the games, the service continued to provide news, sports, weather, and classified advertisements.

Time, Incorporated experimented with teletext but decided not to attempt a full-fledged effort in the commercial teletext business.[11]

Rationales for Regulating Electronic Publication

Electronic publication refers to the various electronic methods of distributing news and information from a "publisher" to an "audience." Publishing has progressed significantly from the hand-operated printing press of the 1700's to the computer-generated, satellite-delivered newspapers of the 1980's. Print publications have traditionally enjoyed the full range of First Amendment protections. But electronic publishing, such as teletext and videotex, obviously did not exist when that Amendment was adopted in 1791, and although they have clearly come of age, as one commentator has noted, they "do not exist in the eyes of the law."[12]

This study focuses primarily on the two newest forms of electronic publication, teletext and videotex. At the present time there is very little direct information on government regulation of these technologies. Congress rarely focuses its attention on electronic publishing, choosing instead to concentrate its energies on telephone, radio, television, and cable television.

Teletext has been the subject of an exhaustive FCC inquiry and report (which will be discussed in-depth in a later section), and the focus of a court challenge, which recently concluded when the Supreme Court let stand a decision of the United States Court of Appeals that allowed the FCC to exempt teletext from content regulation. Videotex, however, remains a virtual regulatory unknown. Services like CompuServe, the

Source, the Dow Jones Retrieval Service, and others operate free from government censorship, but the reason for that freedom could simply be that regulators have not focused attention on this new breed of electronic communicators.

The major public policy question considered in this study is whether existing content controls applicable to traditional over-the-air broadcasting and arguably to cable television should apply to new technologies, such as teletext and videotex, which utilize either the electromagnetic frequency spectrum, cable, or telephone wires to transmit information to the public. A critical factor in the regulatory equation is that neither teletext nor videotex can be delivered without the help of regulated technologies. Teletext is carried over the airwaves on the vertical blanking interval of over-the-air television signals and is deciphered in the home by a specially manufactured decoding device; videotex is transmitted into the home via telephone or cable television wires and is displayed on a computer terminal. Because of this electronic piggybacking, each of these technologies is potentially subject to the regulations now applied to their host technology.

These regulations will be discussed in detail in a later section, but for now, it is important to recognize how the blurring of print and electronic media raises serious policy and constitutional questions. If the government can require television broadcasters to cover controversial issues and offer time for contrasting views, can it similarly require teletext operators who disseminate news and information via the vertical blanking interval of the television signal to be "fair" in their presentations and force these operators to provide free space for replies? Moreover, if videotex is distributed via cable television wires, does the information in the service's database have to be balanced and fair?

This study will examine those rules that regulate the content of electronic communications and discuss which of these are arguably applicable to teletext and videotex. In its analysis section, the report will review arguments over the applicability of these regulations. Before this is done, however, this report will detail the policies behind the adoption of the Communications Act, an Act which neither mentions teletext nor videotex, but which nevertheless arguably controls their development and operation.

A careful understanding of the regulatory rationales for radio and television, cable television, and common carriers is essential before examining the policy and constitutional questions concerning the newest forms of electronic publication.

The History of Radio and Television Regulation

The advent of radio was the spark that ignited the communications revolution. Initial experiments confirming the existence of radio waves were conducted by Heinrich Rudolph Hertz, a German physicist. Shortly thereafter, Guglielmo Marconi accomplished the first successful transmission of radio signals in 1895. Lee DeForest, an American inventor, pioneered the radio vacuum tube. His audion, developed in 1921, made wireless communication practicable. The first high-powered naval radio stations were designed and installed by him.

As the availability of radio transmission devices increased, more amateurs began experimenting, and the electromagnetic frequency spectrum became congested with competing voices. Confusion and chaos ruled the airwaves. The United States Navy, which relied on radio for ship-to-ship and ship-to-shore transmissions, was concerned about the growing interference problem and asked Congress to intervene and prescribe standards for radio use.

The first congressional action on the use of radio was the Wireless Ship Act of 1910, which protected the safety of passengers on ocean-going steamships. Two years later, Congress enacted the Radio Communications Act giving the Secretary of Commerce and Labor the power to license radio stations and operators.

Radio technology continued to develop at a rapid pace, prompting the government to initiate a series of National Radio Conferences to analyze and propose solutions to the problem of station interference. At the first National Radio Conference, Secretary of Commerce Herbert Hoover analogized radio to an important natural resource and urged the adoption of government regulations to protect radio's technological integrity. To his dismay, the Secretary soon discovered that the Radio Act left him powerless to control the number of radio licensees.[13] At the second Radio Conference, there was a proposal to allow the Secretary to alleviate spectrum congestion problems by establishing hours of operation and assigning specific wavelengths to radio operators. A fourth Radio Conference in 1925 produced a resolution for limiting the time and power of individual operators. The courts ruled, however, that the 1912 Radio Act failed to delegate these powers to the Secretary,[14] and therefore, radio operators were left virtually unregulated—free to change frequencies and increase their power and times of operation at will.

Regulations Adopted for Technical Reasons

The broadcasting environment in the late 1920's consisted of stations operating according to their own sets of rules, at the power, time, and wavelengths of their choice. To remedy this situation, President Coolidge asked Congress to adopt a set of federal standards for radio, and Congress answered with the Dill-White Radio Act of 1927.

The 1927 Radio Act created a five-member Federal Radio Commission to assign wavelengths, categorize radio stations, limit power levels, and grant, renew, and deny broadcast licenses.[15] Government licensing decisions were based on a "public interest, convenience, and necessity" standard,[16] a standard which continues to this day. (In the following section, this report will examine how this standard has been used to regulate the content of radio communications.)

Licensing was the government's solution to the interference problem. By assigning radio stations to particular frequencies in the electromagnetic spectrum, the Federal Radio Commission was of necessity limiting the number of potential radio operators. The alternative, however, was anarchy on the airwaves. As we shall see, the government's licensing scheme, in addition to maintaining the technological integrity of the spectrum, saddled licensees with regulations that had nothing to do with technical necessity. For example, the Act mandated that equal opportunities be provided to candidates for political office.[17] Adopted because of fears that licensees would monopolize the airwaves and show political favoritism, this provision still obligates broadcasters to carry the uncensored messages of political candidates.

The Radio Act of 1927 was short-lived. After seven years, Congress replaced it with a more comprehensive set of regulations embodied in the Communications Act, which created the Federal Communications Commission. The FCC was given the broader power of regulating all interstate and foreign communication by wire and radio, including telegraph, telephone, and broadcast. The FCC has the power to grant, suspend, and revoke licenses. The 1934 Act also permitted the FCC to "make such rules and regulations and prescribe such restrictions and conditions, not inconsistent with law, as may be necessary to carry out the provisions of this Act...."[18]

On the purely technical level, the FCC has adopted regulations for dividing the spectrum into different areas and assigning space to different services. The FCC also classifies broadcast stations based on their operating power and distance levels. Other regulations maintained for technical reasons include designating broadcast channels by numbers, allotting and assigning stations, establishing minimum distance separations between stations, and regulating antenna height and power. To monitor compliance, the FCC can require stations to maintain detailed records concerning transmissions and energy output levels, programming, communications, or signals.[19]

When the Communications Act was adopted, the electronic communications landscape consisted of two distinct forms of communication: over-the-air transmission, namely broadcasting, and communication by wire, namely common carriage. The Act defines "radio communication" as "the transmission by radio of writing, signs, signals, pictures, and sounds of all kinds, including all instrumentalities, facilities, apparatus, and services (among other things, the receipt, forwarding, and delivery of communications) incidental to such transmission."[20] Television falls within the definition of radio broadcasting. Common carriage generally refers to telephone communication.[21] According to the Act, these two activities are mutually exclusive, i.e., "a person engaged in radio broadcasting shall not...be deemed a common carrier." Common carriers are regulated under Title II of the Act (content regulation of common carriers will be discussed later in this report), while broadcasting is regulated under Title III. Each Title contains a vastly different set of regulatory controls. *The distinction between hybrid services using broadcast delivery techniques and those using common carrier technology, therefore, is vitally important for postulating the regulatory regime applicable to any new unregulated means of electronic publication.*

Broadcast Regulations Based on The Scarcity of Frequencies

The Fairness Doctrine

The Communications Act of 1934 in sections 315 and 312(a)(7) imposes several government restrictions on news and editorial programming and on political appearances and advertising. While these restrictions will be examined in due course, this report focuses most of its attention on the fairness doctrine because it is by far the most burdensome of all content controls, and it poses the greatest danger to the development of new technologies.

A few years ago, James Batten, the president of Knight-Ridder Newspapers, a company whose Viewtron service was one of the first major publishing efforts in the videotex field, described the potentially disastrous effect of applying the fairness doctrine to developing technologies. Before the Senate Commerce Committee, he said: "The prospect of government regulation to enforce fairness and balance and accuracy is obviously a chilling prospect, and one that could stifle the development of all these new technologies."[22]

The fairness doctrine comes in two parts. The first requires broadcasters to devote a significant amount of time to controversial issues of public importance. The second requires broadcasters to provide a reasonable opportunity for contrasting views on those issues. In theory, the doctrine advances the twin policy goals of keeping the public informed and tempering the views of individual broadcast licensees. In reality, it forces licensees to adhere to a governmentally imposed code of ethics that would be unconstitutional if applied to the print media.

Development of the Fairness Standard

Although the doctrine was first officially expressed in a 1949 Report of the Federal Communications Commission entitled "Editorializing by Broadcast Licensees," the policy it expresses dates back to the days of the Federal Radio Commission. In the congressional debate over the 1927 Radio Act, some members were concerned that the Act would allow the federal government to place restrictions on the content of radio programming. Congressman Fiorello LaGuardia was afraid the Secretary of Commerce would acquire a power "akin to censorship." An exchange between LaGuardia and Representative White clarified the legislative intent of the Act:

> *White:* ...The pending bill gives the Secretary no power of interfering with freedom of speech in any degree.
>
> *LaGuardia:* Is it the belief of the gentleman and the intent of the Congress in passing this bill not to give the Secretary any power whatever in that respect [program control] in considering a license or the revocation of a license?
>
> *White:* No power at all![23]

But not long afterward, in 1929, the FRC evaluated content of broadcasting to determine whether to issue a license. The Chicago Federation of Labor had asked for an increase in hours and in power. The FRC denied the request, saying in part, "all stations should cater to the general public and serve the *public interest* as against group or class interest."[24] The decision was upheld by the Court of Appeals, but it was not reviewed by the United States Supreme Court.

In 1931, the FRC for the first time denied a license renewal based on programming content. A Kansas doctor who also owned a drug company was using his station to promote his hospital, and to prescribe and sell his products over the air. The FRC ordered Dr. Brinkley off the air, declaring his request for license renewal not to be in the "public interest."[25] The denial was upheld by the Circuit Court, and there was no Supreme Court review.

Another denial of license renewal based on program content occurred in 1932 to Trinity Methodist Church, whose station was owned by Reverend Schuler. Over the air, Dr. Schuler expressed personal views on many subjects, including criminal trials, prostitution, Catholics, and Jews. Again, the Court of Appeals upheld the renewal denial and justified the FRC's action as part of its obligation to maintain the public interest.

In 1934, the Radio Act was expanded into the Communications Act, under which the FRC became the Federal Communications Commission (FCC), an independent agency. Although more comprehensive, the 1934 Act replicated most of the terms and conditions of the prior Radio Act.

The first Supreme Court statement on the legality of regulation of broadcast licenses came in *National Broadcasting Company v. United States*.[26] In this case, the major radio network challenged the authority of the FCC to adopt rules concerning network broadcasting. The Commission had established rules which it deemed in the public interest and intended to foster greater program diversity. In a five-year investigation, the FCC found that the National Broadcasting System single-handedly programmed more than 86% of the total nighttime broadcasting signals. The Commission expressed a concern that such concentration of power might adversely affect the ability of licensees to provide service to their listeners.

The Supreme Court ruled against NBC. The Court argued that radio waves were a scarce commodity with a limited opportunity for numbers of licensees. It contrasted that situation with the number of newspapers in existence and the capacity of any new potential publisher to put out an additional newspaper and compete with the existing ones. From that factual distinction, the Court concluded that "freedom of the press" as it appears in the First Amendment, and which clearly forbids any control of the content of the newspapers by the government, did not and should not apply to radio broadcasting.[27] Justice Felix Frankfurter wrote for the majority:

> The act itself establishes that the Commission's powers are not limited to the engineering and technical aspects of regulation of radio communication. Yet we are asked to regard the Commission as a kind of traffic officer, policing the wave lengths to prevent stations from interfering with each other. But the Act does not restrict the Commission merely to

> supervision of the traffic. It puts upon the Commission the burden of determining the composition of that traffic.[28]

This decision, based on the alleged scarcity of broadcast frequencies, is the cornerstone of all later decisions endorsing FCC controls over radio and television broadcasting, including the fairness doctrine.

In 1949, the FCC officially stated its philosophy of fairness in a Report entitled "Editorializing by Broadcast Licensees."[29] In that Report, the FCC declared:

> ...the needs and interests of the general public with respect to programs devoted to news commentary and opinion can only be satisfied by making available to them for their consideration and acceptance or rejection, of varying and conflicting views held by responsible elements of the community. And it is in the light of these basic concepts that the problems of insuring fairness in the presentation of news and opinion and the place in such a picture of any expression of the views of the station licensee as such must be considered.[30]

The Commission's 1949 Report outlined and discussed the obligations of broadcasters to "devote a reasonable percentage of their broadcast time to the presentation of news and programs devoted to the consideration and discussion of public issues of interest in the community served by the particular station."[31] The Report then cited a series of cases in which the Commission had specifically stated that in the presentation of news and public affairs "the public interest requires that the licensee must operate on a basis of overall fairness, making his facilities available for the expression of the contrasting views of all responsible elements in the community on the various issues which arise."

On the issue of licensee editorializing, the Commission, while accepting the contention that "the public has less to fear from the open partisan than from the covert propagandist," rejected the idea that broadcasters should be free to editorialize without having to provide time for an opposing view. The Commission summarily discarded the discussion of broadcasters taking positions on the issues by stating "a licensee should occupy the position of an impartial umpire...we cannot see how the open espousal of one point of view by the licensee should necessarily prevent him from affording a fair opportunity for the presentation of contrary positions or make more difficult the enforcement of the statutory standard of fairness upon any licensee."

In the Report's conclusion, the Commission reiterated what it considered to be a basic policy of fairness:

> [T]he Commission believes that under the American system of broadcasting the individual licensees of radio stations have the responsibility for determining the specific program material to be broadcast over their stations. This choice, however, must be exercised in a manner consistent with the basic policy of the Congress that radio be maintained as a medium of free speech for the general public as a whole rather than as an outlet for the purely personal or private interests of the licensee. This requires that licensees devote a reasonable percentage of their broadcasting time to the discussion of public issues of interest in the community served by their stations and that such programs be designed so that the public has a reasonable opportunity to hear different opposing positions on the public issues of interest and importance in the community.[32]

The 1959 Amendment to the Communications Act

In 1959, Congress amended section 315 of the Communications Act to exempt bona fide newscasts, news interviews, news documentaries, and on-the-spot coverage of news events from the Act's "equal opportunities" requirements (discussed in a later section). In amending the Act, Congress declared that it was not "relieving broadcasters," when covering the exempt events, "from the obligations imposed upon them under this Act to operate in the public interest and to afford reasonable opportunity for the discussion of conflicting views on issues of public importance."

The language of the 1959 amendment reflected a belief on the part of the Congress that the Communications Act *already* imposed a positive obligation upon broadcasters to act fairly and in the public interest. To this day, however, there is still considerable disagreement over whether Congress codified the fairness doctrine in 1959. (The controversy is discussed at the end of this section.)

The Red Lion Decision Upholding the Constitutionality of the Fairness Doctrine

The Supreme Court upheld the constitutionality of the fairness doctrine in 1969 in *Red Lion Broadcasting Co. v. Federal Communications Commission*.[33] The case concerned a one-kilowatt radio station, WGCB, whose signal was broadcast and principally received in Red Lion, Pennsylvania, a town of 5,684 persons. Although the station was the only one physically located in Red Lion, it was in active competition with at least 30 other radio stations readily received by local residents. And 2,080 homes in Red Lion received 12 television stations via cable.

In 1964, WGCB carried a syndicated commentary of Billy James Hargis, known as "The Christian Crusade," in which Hargis attacked a nonresident author named Fred Cook. Cook demanded free time under the fairness doctrine. Red Lion Broadcasting's commercial rate for one hour of prime time was $25. It would have cost Cook only $5 to purchase response time, but he insisted on free time. When the station refused to provide it, the Federal Communications Commission intervened and directed Red Lion Broadcasting to make the time available to Cook. The Supreme Court upheld the FCC's decision, and ruled that the fairness doctrine did not violate the constitutional rights of broadcasters. The Court distinguished between broadcasting and the print media by arguing that the airwaves were scarce, thereby making the medium unique. Indeed, the Court held that the electronic nature of broadcasting singled it out for regulation. The Court said:

> Because of the scarcity of radio frequencies, the Government is permitted to put restraints on licensees in favor of others whose views should be expressed on this unique medium.... It is the right of the viewers and listeners, not the right of the broadcasters which is paramount.[34]

Recent Attacks on the Doctrine's Constitutionality

The doctrine's constitutionality has been subject to recent attack in the courts and at the FCC.[35] In a 1984 case, *FCC v. League of Women Voters of California*, the Supreme Court suggested its willingness to reconsider the validity of the scarcity rationale if Congress or the FCC could show that "technological developments have advanced so far that some revision of the system of broadcast regulation may be required."[36] More recently, in *Telecommunications Research and Action Center v. Federal Communications Commission*, the United States Court of Appeals for the District of Columbia Circuit held that the fairness doctrine was *not* codified in 1959, and therefore, it was never a binding statutory obligation.[37] On June 8, 1987 the Supreme Court denied a petition for certiorari in the *TRAC* case, thus making the decision final. Accordingly, the FCC may now repeal the doctrine if it no longer serves the public interest or is unconstitutional. The FCC has made it very clear that it believes the doctrine is unconstitutional and violates the public interest.[38]

The FCC's position on scarcity indeed answers the Supreme Court's call in *League of Women Voters* for evidence of the rationale's lack of validity:

> [I]n light of the substantial increase in the number and types of information sources, we believe that the artificial mechanism of interjecting the government into the affirmative role of overseeing the content of speech is unnecessary to vindicate the interest of the public in obtaining access to the marketplace of ideas.[39]

Recent Congressional Action To Codify the Fairness Doctrine

In response to the FCC's position on the fairness doctrine, Congress recently passed legislation to codify the fairness doctrine and to prevent the FCC from accomplishing an administrative repeal. This legislation passed the Senate on April 21, 1987 by a vote of 59 to 31.[40] It also passed the House on June 3, 1987 by a vote of 302 to 102. But this legislation was vetoed by President Reagan on June 20, 1987, who cited its infringement on "freedom of expression." Said the President, "This type of content-based regulation by the federal government is, in my judgment, antagonistic to the freedom of expression guaranteed" by the Constitution. The President made it clear that the Supreme Court had consistently declared "intrusion into the function of the editorial process" of the print media unconstitutional. "In any other medium ...such federal policing of the editorial judgment of journalists would be unthinkable....The framers of the First Amendment, confident that public debate would be freer and healthier without the kind of interference represented by the 'Fairness Doctrine,' chose to forbid such regulations in the clearest terms."

No attempt to override the President's veto was made by the Senate. However, Senator Ernest Hollings, Chairman of the Commerce Committee, vowed to tack his codification wording onto another "veto-proof" bill at a later date.

At this time, the fairness doctrine is also in a state of legal limbo. It continues to be enforced by the FCC despite the recent ruling in the *TRAC* case that it is not a binding statutory obligation. If, however, the FCC repeals the doctrine, or if the courts decide that the doctrine's application to radio and television is unconstitutional, then it could not be applied to hybrid technologies, such as teletext and videotex. The FCC has initiated a rulemaking to determine whether enforcement of the doctrine is contrary to the public interest. It acted after the U.S. Court of Appeals in Washington remanded the case in which Meredith Corporation challenged the constitutionality of the doctrine. Action is expected by the end of 1987, but will be preceded by the FCC fulfilling its obligation to Congress to investigate "alternatives" to the fairness doctrine. The "alternatives" report is to be submitted to Congress by September 30, 1987.

Other Broadcast Regulations Based on Scarcity

The Personal Attack and Political Editorializing Rules
These rules are based on the fairness doctrine. The personal attack rule provides that when an attack is made on the honesty, character, integrity, or similar personal qualities of an identified person or group during a presentation of views on a controversial issue of public importance, those who were attacked must be given notice, a transcript, and an opportunity to respond. The political editorializing rule requires broadcasters who endorse candidates to notify opposing candidates and offer reply time.

Equal Opportunities Rule
Section 315 of the Communications Act of 1934 states that a broadcast licensee who permits a legally qualified candidate to use a broadcast station must provide "equal opportunities" to all other legally qualified candidates for that same office to use the station. The rule applies to elections on all levels—local, state, and federal. Moreover, it applies to all legally qualified candidates, including third party and bona fide write-in candidates.

Like the fairness doctrine, the equal opportunities rule is justified by the alleged scarcity of broadcast frequencies.

Reasonable Access By Federal Candidates
Section 312(a)(7) of the Communications Act states that any station's license may be revoked "for willful or repeated failure to allow reasonable access to...the use of a broadcasting station by a legally qualified candidate for Federal elective office on behalf of his candidacy." Licensees must provide broadcast time to any federal candidate who requests it.

Broadcast Regulations Based on Pervasiveness

Obscene and Indecent Broadcasts

Federal law prohibits broadcast stations from using "any obscene, indecent, or profane language by means of radio communication."[41] The rules on indecency are based not on the scarcity of broadcast frequencies, but on the pervasiveness of the medium and its unique accessibility to children and its intrusiveness on nonconsenting adults. Historically, FCC enforcement in this area has been extremely limited, but three recent rulings indicate that it plans on reasserting content control to "protect children" in the viewing and listening audience.

The First Amendment has never protected obscenity, be it printed or broadcast. In Miller v. California,[42] the Supreme Court said communication was "obscene" if it met all of the following tests: "(a) whether 'the average person, applying contemporary community standards' would find that the work, taken as a whole, appeals to the prurient interest . . . ; (b) whether the work depicts or describes, in a patently offensive way, sexual conduct specifically defined by the applicable state law; and (c) whether the work, taken as a whole, lacks serious literary, artistic, political, or scientific value. . . ." Any broadcast that contains obscene speech could subject the broadcaster to criminal liability—a $10,000 fine and/or two years imprisonment.

The Pacifica Decision: "Seven Dirty Words"

In a ruling peculiar to broadcast communications, the Supreme Court extended the class of unprotected expression to indecent speech. In

FCC v. Pacifica Foundation,[43] the Court held that the FCC could impose administrative sanctions upon a radio licensee for broadcasting indecent material at a time when children were likely to be in the audience. In upholding the FCC's decision, the Court said that "of all forms of communication, it is broadcasting that has received the most limited First Amendment protection."[44] The Court noted that broadcasting had a "pervasive" presence and was uniquely accessible to children. The Court split 5 to 4 with Justices Powell and Blackmun concurring.

Pacifica involved a radio station broadcasting 12 minutes of a George Carlin comedy monologue at 2 o'clock in the afternoon. A listener wrote a letter to the FCC saying that he heard the broadcast while driving with his son and was offended by the language Carlin used. Subsequently, the FCC imposed administrative sanctions on Pacifica Foundation's radio station, WBAI, for broadcasting language which it considered "indecent."

In upholding the FCC, the ruling plurality of the Supreme Court focused on the *pervasiveness* of the broadcast medium, and two concerns flowing from that: the protection of children from socially undesirable but otherwise protected speech, and the intrusiveness of such speech into the private homes of adults who find such material offensive.

Recent FCC Rulings on Indecency

As stated earlier, the FCC recently released a public notice containing new indecency enforcement standards to be applied to all broadcast and amateur radio licensees. The Commission announced that it would no longer limit indecency to the actual seven dirty words in the Carlin monologue. Instead it would use the following generic definition of broadcast indecency: "language or material that depicts or describes, in terms patently offensive as measured by contemporary community standards for the broadcast medium, sexual or excretory activities or organs."[45] Using this definition, the FCC found that a broadcast of excerpts from a play entitled "The Jerker" on KPFK-FM (Los Angeles, California) radio, which was broadcast at 10 p.m. and was preceded by a warning as to content, was indecent.[46] It also ruled that various episodes of the Howard Stern radio program broadcast from New York City "may give rise to actionable indecency." Moreover, the broadcast of the song "Makin' Bacon," which contained a number of references to sexual organs and activities, was found indecent. Although the material was broadcast after 10 p.m., the FCC stated that "there was still a reasonable risk that children may have been in the audience at that time."[47]

Regulation of Cable Television

Cable television was launched in the late 1940's as a method of providing television signals to communities where normal television reception was poor or nonexistent. To receive over-the-air signals, communities mounted large antennas on nearby hilltops and transmitted the signals via cable into local homes.

The principal source of governmental regulation over cable television, from its earliest days to the present, has been the need to obtain some form of permission to install wires along, over, or under various public rights of way in order to connect subscribers to the cable system. Before granting permission, local governments set conditions and extract commitments through a franchise auction process.

Regulation at the federal level has been premised on the assumption that cable systems operate in interstate commerce. Until recently, the FCC's jurisdiction over cable television was viewed as legitimate as long as it remained ancillary to the agency's regulation of broadcasting.[48] Because cable systems have been held to engage in interstate communication by wire, they could also be regulated as common carriers to the extent that the services they provide constitute common carrier services. In *National Association of Regulatory Utility Commissioners*, the court of appeals held that a cable system could be subject to common carrier regulation on certain non-video channels, if, in fact, the cable system operated those channels in a manner which otherwise met the defining characteristics of common carriage.[49]

Application of the Fairness Doctrine and
Equal Opportunities Rule to Cable

The FCC has imposed certain broadcast regulations on cable systems, often without justification. After the Supreme Court ruled that regulation by the FCC of cable was justified as long as it was "reasonably ancillary" to its broadcast regulation responsibilities, the FCC extended the fairness doctrine and equal time requirements to cable.[50]

The fairness doctrine applies to "origination cablecasting," that which is under the exclusive control of the cable operator. Most experts believe this rule applies to cable programming that is produced by the local cable company. Thus, the fairness doctrine would not apply to retransmission of local or distant television broadcasts. This point is moot since the fairness doctrine applies at the point of origination in such cases. For example, it applied to WGN in Chicago and to Turner Broadcasting's WTBS in Atlanta, both commonly retransmitted by cable systems.

But does the fairness doctrine apply to programming delivered by satellite such as HBO, Showtime, and the Disney Channel *if* they engage in news and editorial programming? Again, for the most part, experts believe that the doctrine would not apply to such programming. But William Johnson, the acting chief of the Mass Media Bureau at the FCC, says that the application ruling is ambiguous enough to allow a pro-regulatory FCC to apply the fairness doctrine to satellite-delivered programming in the future.

The equal time requirements for cable were statutorily adopted in 1972 in an amendment to Section 315 of the Communications Act. The amendment defines "broadcasting station" to include "a community antenna television system," and "licensee" as the operator of the system. Despite the questionable application of broadcast rules to cable, which does not use the spectrum, cable operators must abide by these rules in the absence of congressional or court action stripping the FCC of the authority to enforce content controls against cable.

Recent Congressional Action on Fairness Doctrine
And Its Applicability to Cable Television

On April 21, 1987, the Senate debated and passed S. 742, the Fairness in Broadcasting Act of 1987, which would have codified the fairness doctrine. The legislative history of the bill makes it clear that the doctrine applies to cable as well as broadcasting, and arguably to any hybrid services that might be carried over cable.

Senator Hollings, a primary cosponsor of the legislation, inserted the following statement into the record during the floor debate:

> Cable television operators are currently obligated to follow the Fairness Doctrine, and S. 742 makes no change in this requirement. While the application of the doctrine to cable operators may appear to cover a wide range of programming, in effect, it applies in only one key respect: locally originated material ('origination cablecasting', See 47 CFR 76.209). Whenever a cable operator originates a local program *or inserts local material into programming originated elsewhere*, he is obligated to comply [with] the doctrine's requirements. [emphasis ours]
>
> The application to cable operators rests on three grounds. First, cable television operators have deliberately integrated their offerings with those of broadcasters. Most of the programming viewed by cable television audiences is retransmitted television signals. When these audiences "turn the dial," they do not differentiate between these retransmitted programs and other offerings. Thus, cable television is a surrogate for broadcasting and, from a public interest viewpoint, needs to follow similar requirements.
>
> Second, it is important that any local broadcasts or locally originated programming uphold the purposes of the Communications Act. This viewpoint was best expressed in United States v. Midwest Video Corp., 406 U.S. 649 (1972), where the Supreme Court upheld the FCC's requirement that cable operators originate local programming if they wish to carry broadcast signals. The rationale for this requirement was that the FCC could oversee cable television "with a view not merely to protect but to promote the objectives for which the Commission had been assigned jurisdiction over broadcasting." Id. at 667.
>
> Third, as with its grant of broadcasting licenses, the Federal Government has established extensive requirements for cable television systems to ensure the public interest is served. These include elaborate rules for the granting of a franchise to use public resources. The 1984 Cable Communications Policy Act (P.L. 98-549), which contains these policies, has a basic purpose to "assure that cable communications provide and are encouraged to provide the widest

possible diversity of information sources and services to the public". Congress viewed the affirmative obligations to enhance diversity as a trade-off for the benefits granted cable operators. Moreover, the Fairness Doctrine obligations are directed only toward locally produced material, a very small part of cable television operations, and are a limited intrusion. In balancing these various interests, it is clear they place a minimal burden on cable television operators while forwarding the interests of the public in receiving a greater diversity of views. See City of Los Angeles v. Preferred Communications, 106 S. Ct. 2034 (1986).

If the contents of S. 742 ever become law, Senator Hollings' statement will undoubtedly become part of its legislative history. (See the earlier section on the fairness doctrine; also see section describing how President Reagan vetoed the Fairness in Broadcasting Act of 1987 on June 20, 1987.) Absent a successful court challenge to this provision, there would be no question that the fairness doctrine and other content controls are applicable to locally produced cable programming. Videotex services transmitted by wire and targeted at the local community might also be subject to these restrictions.

Obscene and Indecent Cable Programming

Cable programming that is obscene is not protected by the First Amendment. Section 639 of the Cable Communications Policy Act of 1984 imposes criminal sanctions on persons transmitting "over any cable system...any matter which is obscene or otherwise unprotected by the Constitution." The legislative history of the Act adopts the Supreme Court's *Miller v. California* definition of obscenity, which was discussed in the section on broadcast obscenity. While *Miller* would arguably permit regulation of cable programming that is obscene, it does not allow the federal government or the state governments to regulate "indecent or profane material" as the federal government may now regulate over-the-air broadcasting.[51] Programming that is merely "indecent" does not fall within the bounds of *Miller,* and the Supreme Court recently upheld a decision by the Tenth Circuit Court of Appeals which prohibited the State of Utah from regulating indecent cable programming.[52]

Regulation of Common Carriers

In General, No Content Regulation

Common carriers, by their nature as carriers for hire, cannot exercise editorial control over the messages transmitted over their systems. Indeed, the Communications Act requires that a telephone company provide access to its facilities "upon reasonable request." The carrier has no right to deny the use of its facilities to anyone. Therefore, common carriers cannot interfere with the content of the messages they transmit. Similarly, the government has no power to control the content of telephone conversations, except in the limited areas of obscenity, dial-a-porn services, and harassing and annoying calls.

Obscenity, Indecency and Abusive Calls

Federal law prohibits anyone from making obscene or harassing telephone calls in interstate or foreign communication by means of telephone. Additionally, most states regulate this type of behavior.

Section 223(a) of the Communications Act makes it unlawful to "make any comment, request, suggestion, or proposal which is obscene, lewd, lascivious, filthy, or indecent." It also prohibits harassing or threatening calls. Section 223(b) provides:

> Whoever knowingly...by means of telephone, makes (directly or by recording device) any obscene or indecent

communication for commercial purposes to any person under eighteen years of age or to any other person without that person's consent, regardless of whether the maker of such communication placed the call; or permits any telephone facility under such person's control to be used for an activity prohibited by subparagraph (A), shall be fined not more than $50,000 or imprisoned not more than six months, or both.

The section also adds that "it is a defense to a prosecution under this subsection that the defendant restricted access to the prohibited communication to persons eighteen years of age or older in accordance with procedures which the Commission shall prescribe by regulation."

The most recent congressional activity in this area concerned so-called "dial-a-porn" services. Dial-a-porn is a "dial-it" service which allows multiple callers to simultaneously access prerecorded messages. To access a dial-a-porn service, the caller dials a phone number and hears a description or depiction of actual or simulated sexual behavior. The messages are typically changed on a daily basis. The common carrier does not operate the message service but provides the line pursuant to an *intrastate* tariff. These tariffs generally provide that the subscriber has exclusive control over the content and quality of the messages recorded and that the telephone company assumes no liability.

Congress focused its attention on dial-a-porn services after constituents complained that the service was accessible to children. During Congress's inquiry, the FCC ruled that the Communications Act's prohibitions on obscene, lewd and indecent telephone calls did not apply to dial-a-porn because section 223 was originally intended to apply to calls that were deliberately made to innocent, nonconsenting individuals.[53] In the aftermath of this ruling, Congress amended section 223 and expressly brought dial-a-porn services within the statutory prohibition.

Under the amendment, a person is liable for violating section 223 regardless of who places the call. The amendment applies only to interstate calls or calls within the District of Columbia. The amendment is targeted at the availability of obscene material to children. On this point, Congressman Bliley noted: "The inability of parents to effectively control children's access to obscene material places a duty on the government to intervene."[54] The new section 223 is targeted at providers of dial-a-porn material and not at the common carriers, who are by law prohibited from listening to or affecting the content of telephone conversations. The prohibitions apply to all commercial enterprises that use a common carrier to disseminate prohibited material. Although the law has yet to be applied as such, it could be interpreted to prohibit videotex services from making sexually suggestive material available to an audience of users that might include unsupervised children.

Application of Broadcast Content Controls to Teletext

Telecommunications Research and Action Center v. FCC

On September 19, 1986, the United States Court of Appeals for the District of Columbia Circuit announced its decision in *Telecommunications Research and Action Center v. FCC*.[55] The case was brought to challenge the FCC's decision *not* to apply its content regulations, namely the fairness doctrine, equal opportunities and reasonable access rules, to teletext. The Court's decision, written by Judge Robert Bork and joined by Judge Antonin Scalia, upheld the FCC's decision, but it also firmly established that the FCC's authority to regulate the content of radio, television, and cable television signals extends to hybrid technologies *using the broadcast spectrum*.[56]

While the decision provides regulatory relief for teletext entrepreneurs, the permanence of such relief depends in large part on the regulatory philosophies of future commissioners. Moreover, the *TRAC* court noted that the question of future application of content regulations could be mooted if these restraints, based on scarcity of the electromagnetic frequency spectrum, were found to be unconstitutional.[57] Nevertheless, *TRAC* is the most important decision to be issued concerning the extension of content regulation to new communications technologies. It is also important to note that TRAC's petition for certiorari with the

Supreme Court has been denied, leaving the Appellate Court's decision as regnant law.

FCC Authorization of Teletext

The FCC's investigation of teletext began on October 22, 1981 when it adopted a *Notice of Proposed Rulemaking* to consider whether it was in the public interest to authorize television licensees to operate teletext systems.[58] The *Notice* was issued at the request of station operators who expressed considerable interest in providing teletext services. Initially, the Commission sought input on how best "to provide a regulatory environment that is conducive to the emergence and implementation of new technology and new uses of the [broadcast] spectrum."[59] Regulation was viewed as a potential impediment to the development of teletext services; the Commission expressed its collective belief that "the forces of competition and the open market" would better serve the public's interest in receiving information from a diverse marketplace of ideas.[60]

Subsequently, the Commission released a *Report and Order* containing specific rules authorizing licensees of both low power and full power television stations to operate teletext services.[61] The *Report* granted station licensees complete editorial control over teletext services and deferred the decisions concerning the technical specifications of the hardware for transmitting and receiving teletext signals to marketplace forces. The only limitations the FCC placed on teletext operators were that "teletext operations must not interfere with the regular broadcast service of the originating station, the signals of other broadcast stations, or the signals of non-broadcast radio stations."

In its *Report* the Commission also decided that teletext should not be regulated like broadcasting, but rather, it should be treated "as an ancillary broadcast service." This meant that content regulations would be enforced against the station's primary broadcast service, the broadband signal carrying video programming, but not against the ancillary service, the teletext signal transmitting written pages and hard data.[62] The Commission decided that its political broadcasting regulations, applicable to radio, television, and cable television, "as a matter of law, ...need not be applied to teletext service," and furthermore "teletext's unique characteristics" as a hybrid print and broadcasting medium made application of the content restrictions "unwise as a matter of policy."

The FCC articulated four justifications for what it characterized as "the open regulatory approach." First, the FCC found many potential uses for teletext which were incapable of orderly characterization. Second, the FCC noted that potential applications for teletext services would "vary across different markets and user groups." The FCC described three different markets: one offering an advertiser-supported consumer service, another providing a restricted access subscriber service, and a third offering business data services. Third, the FCC concluded that the pattern of demand for teletext services would "shift over time as user interests change and new types of services are conceived and implemented." Fourth, the FCC provided a brief regulatory cost-benefit analysis for its hands-off approach: "[I]t would be very costly and difficult for the government to attempt to develop and apply specific performance regulations and standards to meet the wide variety and changing needs of teletext service."

In its analysis of the application of each separate content restriction to teletext, the Commission first held that section 312(a)(7)'s requirement of allowing reasonable access to legally qualified federal candidates "is adequately satisfied by permitting federal candidates access to a licensee's regular broadcast operation; it does not require access to ancillary or subsidiary service offerings like teletext." Their justification was that "teletext offerings [would] not provide a candidate access to the broad television audience attracted to the station's regular broadcast operation." The Commission also held that section 315 was "not applicable to teletext offerings" because teletext was "inherently not a medium by which a candidate can make a personal appearance" necessary to trigger the equal opportunity requirement. Finally, the Commission held that the fairness doctrine should not apply to teletext:

> [T]he 1959 legislative enactment concerning the Fairness Doctrine in Section 315 does not mandate extension of the Fairness Doctrine to new services like teletext, which did not even exist at the time when Congress acted. Rather, any determination concerning this question is one which has been entrusted to our sound judgment and discretion in the first instance. Indeed, in light of the important First Amendment considerations raised by application of the Fairness Doctrine to new communications services, we are reluctant to extend these policies unless we find there is a compelling reason to do so.

The Commission argued that "teletext's unique blending of the print medium with radio technology fundamentally distinguishes it from traditional broadcast programming." Indeed, the Commission contended that teletext more closely resembled print communication. The test for regulating the content of teletext boiled down to requiring "a substantial showing that teletext, solely because it utilizes a different method of delivery to readers, is likely to overshadow all other print sources of information." The FCC concluded that there was no basis in the record for such a finding.

The FCC also accepted the argument that new communications services might not be economically viable if licensees were burdened with fairness doctrine obligations:

> We are persuaded that the likelihood of licensees' embarking upon these types of endeavors will be substantially affected by our determination to apply, or not to apply, traditional broadcast policies like the Fairness Doctrine. We have no desire to block from the outset full development of this promising new service by the unreflective application of requirements that appear fundamentally unsuitable and which are not legally required. Such a course would be inconsistent with our statutory responsibility to promulgate policies that are responsive to the characteristics of new communications services so as to encourage, not frustrate, their development.[63]

This same argument was raised during the Senate Commerce Committee's 1984 hearing on content controls over the electronic media.

After the FCC terminated its *Notice* and adopted its *Report*, two public interest groups filed motions to reconsider the decision. Media Access Project (MAP) argued that teletext should be treated the same as traditional broadcasting. MAP contended that the Communications Act did not give the FCC the discretion to refuse to apply the fairness doctrine to teletext services. Moreover, MAP argued that the Commission had an obligation to ensure political access to teletext. Henry Geller, Donna Lampert, and Philip Rubin argued that the full panoply of political broadcasting regulations should be applied to teletext. The Commission rejected these petitions,[64] and the case went to the U.S. Court of Appeals for the D.C. Circuit.

The TRAC Decision

The Court of Appeals rejected the FCC's argument that the First Amendment prevented broadcast regulation from being applied to teletext. The Court stated that "teletext is transmitted over broadcast frequencies that the Supreme Court has ruled scarce and this makes teletext's content regulable." Although the Court seriously questioned the validity of the scarcity rationale, it said that "[t]eletext, whatever its similarities to print media, uses broadcast frequencies," and therefore, only a reversal of existing precedent could trigger First Amendment protection for teletext. The Court stated: "The Commission...cannot on first amendment grounds refuse to apply to teletext such regulation as is constitutionally permissible when applied to other, more traditional, broadcast media."

The Court found that teletext fell within the definition of radio communication, and that "the Commission's attempt to distinguish teletext from the traditional broadcast mode of mass communication by calling it an ancillary service...departs without explanation from well-established precedent." The Court also rejected the FCC's finding that teletext was incapable of producing a "use" as required by section 315.

The court agreed with the Commission's decision to exempt teletext from the requirements of the fairness doctrine, but it added that because "[t]eletext is broadcast time operated by Commission licensees or by lessees under the control of licensees," *the "fairness doctrine by its terms applies to teletext" and "no extension" of the doctrine "is necessary."* [Emphasis ours.] The implication of this statement could not be clearer: There is nothing to stop a future Federal Communications Commission from imposing fairness doctrine obligations on teletext operators.

In the most important part of the decision, however, the Court found that the fairness doctrine was merely part of the general public interest standard in the Communications Act and *not a binding statutory obligation.* As such, the Court said: "Because the fairness doctrine derives from the mandate to serve the public interest, the Commission is not bound to adhere to a view of the fairness doctrine that covers teletext." The Court added: "To the extent that the Commission's exemption of teletext amounts to a change in its view of what the public interest requires, however, the Commission has an obligation to acknowledge and justify that change in order to satisfy the demands of reasoned decision-making." The Court endorsed the Commission's justification for its refusal to apply the fairness doctrine to teletext: "that the burdens of applying the fairness doctrine might well impede the development of the new technology and that the 'likelihood of licensees' embarking upon...endeavors [like teletext] will be substantially affected by the agency's policy." The Court also agreed that the Commission's decision was in line with the Communications Act's requirement that the Commission encourage the development of new technologies.

In another important development, the Court concluded that spectrum scarcity was no longer a valid rationale for regulating the content of electronic communications. The Court noted that "the line drawn between the print media and the broadcast media, resting as it does on the physical scarcity of the latter, is a distinction without a difference." Moreover, in the specific context of regulating new technologies, the Court stated: "Employing the scarcity concept as an analytic tool, particularly with respect to new and unforeseen technologies, inevitably leads to strained reasoning and artificial results." The Court contrasted the regulated status of broadcasting with the freedom accorded print journalism:

> It is certainly true that broadcast frequencies are scarce but it is unclear why that fact justifies content regulation of broadcasting in a way that would be intolerable if applied to the editorial process of the print media. All economic goods are scarce, not least the newsprint, ink, delivery trucks, computers, and other resources that go into the production and dissemination of print journalism. Not everyone who wishes to publish a newspaper, or even a pamphlet, may do so. Since scarcity is a universal fact, it can hardly explain regulation in one context and not another. The attempt to use a universal fact as a distinguishing principle necessarily leads to analytical confusion.

With this passage, the Court of Appeals joined the growing number of voices calling for a critical reexamination of the scarcity rationale. Should the Supreme Court adopt this line of reasoning, it would be compelled to reverse *Red Lion*, thereby terminating the application of the fairness doctrine. Since the above passages reflect the views of Justice Scalia, and Justice Brennan opposes the imposition of content controls over the electronic media, it is not unreasonable to assume that at least one conservative and one liberal justice would vote to overturn the 1969 decision.

Implications For Content Control of Teletext

The *TRAC* decision sanctions continued governmental control of the content of teletext transmissions based on the scarcity rationale upheld by the Supreme Court in *Red Lion Broadcasting v. FCC*. Despite the *present* Commission's reluctance to embark upon teletext regulation, the authority exists for a *future* Commission. For this authority to be removed permanently, either Congress would have to pass a statute, or the courts would have to overrule the *Red Lion* decision.

The most significant part of the *TRAC* decision, however, was the court's statement that the fairness doctrine was *not* codified by Congress in 1959, and therefore, it is not mandatory for the FCC to apply its requirements to new communications technologies. The Court of Appeals stated:

> We do not believe that language adopted in 1959 made the fairness doctrine a binding statutory obligation; rather, it ratified the Commission's longstanding position that the public interest standard authorizes the fairness doctrine. The language, by its plain import, neither creates nor imposes any obligation, but seeks to make it clear that the statutory amendment does not affect the fairness doctrine obligation as the Commission had previously applied it.... Because the fairness doctrine derives from the mandate to serve the public interest, the Commission is not bound to adhere to a view of the fairness doctrine that covers teletext.

Thus, the application or non-application of the fairness doctrine to other communications technologies subject to the jurisdiction of the FCC is a judgment call to be made according to the Commission's subjective view of what the public interest requires.

As stated earlier, the Supreme Court refused to hear the appeal of certain groups seeking a reversal of the *TRAC* decision. Thus, the decision is good law. Under present law and FCC regulations, however, the fairness doctrine and equal opportunities rules continue to be applicable to radio and television, and teletext transmissions.

Regulation of Indecent Teletext Material

The ban on "obscene, indecent, and profane" material applies to any "means of radio communication." Since the *TRAC* court found teletext to fall within the definition of radio broadcasting, the prohibition on indecent transmissions is apparently applicable to material delivered via the vertical blanking interval. This question, however, has never been raised, and there are numerous arguments why the indecency rules should not be applied to teletext. First, although teletext is radio communication, it is not as pervasive as the AM and FM radio signals any listener can receive. Unlike the situation in the *Pacifica* case, teletext cannot be received on a car radio. Second, teletext is not pervasive in the sense that it is an intrusive medium. Viewers must *purchase a decoder* and learn to use it before they can receive teletext messages. Parents who want to protect their unsupervised children can simply put the decoder in a secure place.

Application of Broadcast Content Controls to Videotex: An Area of Great Uncertainty

The Communications Act gives the FCC very broad powers over communications services. Before it can regulate a service, however, the FCC must demonstrate that what is involved is "communications," and not some form of data processing outside the scope of FCC jurisdiction.

It can be argued that videotex involves "commerce in communication by wire," which subjects it to FCC jurisdiction under section 2 of the Communications Act. Moreover, the interactive nature of videotex distinguishes it from non-communicative activities. For example, when an individual uses a computer to retrieve information stored in that computer's memory, no communication is involved. By comparison, when that same individual connects a telephone line or cable to that computer enabling it to interact with other computers and draw information from distant databases, then a communications link is established and FCC jurisdiction is triggered.

Videotex is not now subject to any FCC regulations peculiar to it. As stated earlier, it is a regulatory unknown. It is very different from teletext. Videotex does not use the broadcast spectrum as a means of transmission. It is disseminated only by wire. Thus, it cannot be subject to regulations based on the scarcity of broadcast frequencies or the pervasiveness of the broadcast medium.

At the present time, videotex technology can be transmitted to the user by two different methods of delivery: telephone lines and cable television wires. As noted throughout this study, each method of delivery is subject to different rules and regulations, particularly in terms of content regulation. Since videotex is clearly not radio communication, it can only be regulated, if at all, as a cable or telephone service.

Regulation of Cable-Delivered Videotex

Congress passed the Cable Communications Policy Act of 1984 to "establish a national policy concerning cable communications." It was also designed to "establish guidelines for the exercise of Federal, State, and local authority" as it concerns cable television. Most important for the development of new technologies, the Act was written to "assure that cable communications provide and are encouraged to provide the widest possible diversity of information sources and services to the public." Any governmental actions inhibiting the development of cable-delivered videotex services would violate the intent of the Cable Act.

For the purposes of this discussion, a few terms need to be defined. The Cable Act defines "cable service" as "the one-way transmission to subscribers of video programming or other programming service." "Other programming service" is defined as "information that a cable operator makes available to all subscribers generally." "Video programming" means "programming provided by, or generally considered comparable to programming provided by, a television broadcast station." While videotex is a two-way service rather than one way, it might fall within this definition if a cable operator used two channels, one for incoming and one for outgoing messages, to deliver videotex information.

The Act also provides cable channels for public, educational, or governmental use. The Act establishes cable channels for commercial use by persons unaffiliated with the system operator. These "leased access" channels are available for any use, which would include providing videotex services, and the Act sets out requirements for the setting aside of leased access channels.

Provisions dealing with leased access include the following:

> [T]he cable operator shall establish...the price, terms, and conditions of such use which are at least sufficient to assure that such use will not adversely affect the operation, financial condition, or market development of the system....A cable operator shall not exercise any editorial control over

> any video programming provided pursuant to this section, or in any other way consider the content of such programming.

Logically, if the fairness doctrine applies only to origination cablecasting by the system operator, it would not apply to videotex carried over leased access channels that are not under the control of the system operator. Section 612(h) provides:

> Any cable service offered pursuant to this section shall not be provided, or shall be provided subject to conditions, if such cable service in the judgment of the franchising authority is obscene, or is in conflict with community standards in that it is lewd, lascivious, filthy, or indecent or is otherwise unprotected by the Constitution of the United States.

Most likely, the scope of this provision would be limited to obscenity.

As this study has shown, the fairness doctrine applies to programming originated by the cable operator and which is under its exclusive control. It does not apply to material on cable access channels because they are protected by the National Cable Policy Act and local franchise agreements, and programming on them is not under the control of the local cable operator. Therefore, only videotex that is originated by the cable operator would be subject to the fairness doctrine. This includes those instances in which a cable operator originates a local program or inserts local material into programming generated elsewhere.

A videotex service that is merely a means of routing users to different outside databases of distant providers would not likely be subject to fairness doctrine controls. Videotex information on local issues, especially when generated by the cable operator, it could be argued, would be subject to content controls.

Indecent Videotex Material Delivered Via Cable

Indecent videotex material delivered via cable television would not be subject to government control. Only videotex that was obscene could be prohibited.

Videotex does not fall under the "pervasiveness" standard applied in *Pacifica*; it is not an unwanted "pig in the parlor." Videotex is only available to those taking the affirmative step of contacting the cable operator and asking that a wire be brought into the home and connected to their television. To receive cable-delivered videotex services a subscriber must pay a monthly cable fee. As the Supreme Court stated in *Erznoznik v. City of Jacksonville*,[65] the fact that a commercial enterprise directs its programming only to paying customers presumably establishes that those customers are neither unwilling viewers nor offended. *They invite the material into the privacy of their home well aware of its contents.*

Furthermore, the Cable Communications Policy Act requires all cable operators to make available to their subscribers "a device by which the subscriber can prohibit viewing of a particular cable service during periods selected by the subscriber."[66] A parent subscribing to cable-delivered videotex may acquire a "lock box" to prevent reception of certain cable channels, which would include the videotex channel, without his or her authorization. Finally, the cable subscriber can terminate the videotex service at any time simply by informing the cable operator that his subscription should be cancelled. Thus, cable-delivered videotex is by its very nature no more intrusive than any home-delivered newspaper, magazine, book, or record. As such, it should be entitled to full First Amendment protection.

Regulation of Telephone-Delivered Videotex

Content Controls

Under present law, broadcast content controls such as the fairness doctrine and equal opportunities rules do not apply to material delivered over telephone wires. Thus, videotex services received via a modem attachment to a personal computer would not be subject to broadcast content controls.

Obscene and Indecent Videotex Material

The Communications Act of 1934 states that a person can be fined or imprisoned for using the telephone to make "any comment, request, suggestion or proposal which is obscene, lewd, lascivious, filthy, or indecent." This section, originally written to protect people from obscene phone calls, raises a number of problems for videotex service operators using telephone facilities to transmit information. For example, certain material that newspapers are free to print may be unacceptable when carried by videotex. Advertisements for X-rated movies in local theatres or video stores delivered to consumers via videotex may be considered filthy or indecent. Recent congressional action has sufficiently broadened the scope of the indecency prohibitions to subject the provider of such material to criminal liability. (See earlier discussion of dial-a-porn.)

Is it possible that the courts would allow the federal or state governments to regulate the decency of videotex material? The Supreme Court's recent decision prohibiting states from regulating indecent cable programming argues against such a regulatory regime. Moreover, as noted earlier, videotex is not a pervasive medium. It is not transmitted over the airwaves. It cannot be received over a car radio, or any radio for that matter. Videotex enters the home by a wire, with the subscribers having full control over the information they choose to access.

Implications of TRAC Decision for Videotex

Although the *TRAC* decision focused on government control of teletext, it has broader implications for videotex providers. The *TRAC* court defined teletext by excluding the transmission of text and graphics by way of cable or telephone, and it expressly stated that its decision did not affect the FCC's authority to regulate the content of material transmitted via those mediums. Furthermore, Judge Bork (with Judge Scalia concurring) made clear that in the near future the Supreme Court would have to resolve the inequity of applying content controls to the electronic media and not to the print media. Then how would the FCC be expected, if at all, to approach the subject of videotex regulation?

If the FCC has jurisdiction to regulate videotex as communication, what things is it likely to consider in its determination of the public interest as regards videotex? The FCC's teletext *Report*,[67] which was the subject of the *TRAC* case, makes clear that the agency would examine the unique characteristics of each new communications technology. It would look at everything distinguishing videotex from other regulated technologies. Moreover, the FCC's teletext *Report* demonstrates that it would be concerned with the likelihood that entrepreneurs would be inhibited from entering the videotex or electronic publishing market if content regulations were applied. The FCC would also analyze the demand for videotex services, and see if it cut across many different markets. Moreover, it would examine the need for videotex operators to adapt quickly to changing consumer demands. Finally, based on its teletext *Report*, the FCC would likely look at the cost and difficulty of regulating videotex and developing content standards.

Impact of Content Controls

The FCC's 1985 *Fairness Report* found substantial evidence that the fairness doctrine chills the discussion of controversial issues on radio and television,[68] and a similar result can be forecast for videotex and teletext. The FCC's *Report* noted that adhering to fairness requirements involves significant burdens. For example, to abide by the doctrine, broadcasters must discover issues that are both controversial and of public importance in the local community. If a particular issue is not covered, a complaint may be filed with the FCC. The typical fairness doctrine dispute, however, involves the obligation to provide time for contrasting viewpoints on those issues the broadcaster has chosen to cover. Whenever a broadcast raises a controversial issue of public importance, whether it be part of standard programming, news, documentaries, or even advertisements, the potential for a fairness complaint exists. Indeed, the FCC's *Fairness Report* noted that broadcasters realize "the probability that coverage of a highly controversial issue will trigger an avalanche of protests demanding air time for the presentation of opposing viewpoints." All complaints must be taken seriously because the possible sanctions include loss of license.

Sometimes fairness complaints are filed simply to harass and intimidate licensees into censoring certain material. Broadcasters know that one fairness doctrine burden is the litigation expenses necessary to defend editorial decisions. For example, KREM-TV in Spokane,

Washington spent $20,000 in legal fees, 480 hours of executive time, and endured a lengthy delay in its license renewal all because of one editorial it broadcast supporting EXPO-74, the proposed world's fair to be held there. Another example of the significant financial burdens resulting from fairness litigation is the NBC award-winning documentary on pension abuses entitled "Pensions: the Broken Promise." A complaint was filed against NBC alleging the program only presented one side of the pension issue, and the FCC agreed and ordered responsive programming. NBC fought the decision and won, but the battle lasted four years and cost the company over $100,000 in legal expenses.

In addition to editorials and documentaries, issue advertisements are subject to the fairness doctrine. Issue advertisements are paid announcements that advocate ideas instead of selling products. The best example of issue advertisements are the free enterprise ads Mobil Oil places in publications such as *Time* magazine and the *Wall Street Journal*. Station attorneys are regularly called upon to pre-screen issue ads and determine if they will trigger a response.

Just as broadcasters must perform this inquiry, teletext operators offering advertising space would have to determine (1) whether or not the advertisement promotes a product or an issue; (2) if it promotes an issue, whether or not it is controversial in the local community; and (3) if the issue is controversial, whether the ad addresses the issue in a meaningful way. If the answer is "yes" to these three questions, then a response would be triggered under the fairness doctrine. Many broadcasters, for financial or legal reasons, simply adhere to a policy of rejecting all issue advertisements.

Imposition of content controls would inhibit teletext and videotex just as they are trying to develop new markets. Applying the fairness doctrine to teletext would effectively preclude the discussion of controversial issues. The page limitations of the technology would make it impossible to cover all controversial issues in a given community. Moreover, it would be impossible to give space to the broad array of contrasting views. Issue advertisements would be effectively precluded from the service because of their controversial nature. Application of the equal time rule to teletext would force teletext operators to remove candidate advertising from the service and carefully screen all pages for references to candidates. The notification requirements and obligations to provide free reply space would be an enormous administrative burden for the teletext operator. Moreover, the necessity of providing space for contrasting views would force operators to lessen the amount of space reserved for other important consumer information. Application of the indecency standards to teletext would also require the operator to devote significant energies to deciphering the vagaries of the standards.

The application of content controls to videotex is even more problematic. What material would be covered by the fairness doctrine? For broadcasters, the only material covered is that which is finally aired. The information used to produce a documentary, for example, is not subject to fairness controls. But for videotex, would the doctrine apply to every page of information in the provider's database, conceivably hundreds of thousands of pages? If the doctrine were applied that broadly, the videotex operator would have to spend all of his time screening the material. The service operator would have to investigate whether the material discussed any controversial issues of public importance to the local community, whether other material in the database balanced that viewpoint, and, if not, who should be contacted to offer a reply. It would be like telling a librarian to read all the books in the library, decide if the library as a whole is balanced and fair in the minds of the members of the FCC, and, if not, to buy more books.

The material in a videotex database is potentially unlimited, and the number of databases is also unlimited. Application of the equal time rule to videotex would also unleash an enormous administrative burden requiring the videotex operator to pre-screen thousands of pages of information. The same goes for indecency requirements. The only possibility in this area would be to require videotex operators to prohibit dial-a-porn services from using their systems. But again, this would require the videotex operator to exercise an enormous degree of content control over information providers.

Conclusions

(1) Teletext is subject to the same content controls as traditional broadcasting services, but by virtue of an FCC ruling, it is presently exempt from these controls because the current FCC has decided not to apply them. *There is nothing, however, preventing a future FCC from enforcing these controls on teletext, videotex, and a host of other electronic technologies.* The current trend in Congress is to encourage greater regulation.[69]

(2) Teletext may not transmit obscene material, but whether it may transmit indecent material is an open question.

(3) Videotex is not a form of traditional broadcasting. It does not use the airwaves and should not be subject to broadcast content controls. Although cable television is subject to the fairness doctrine and equal opportunities rules, it is unclear what a court would do if faced with an FCC decision applying the fairness doctrine to cable-delivered videotex. While not specifically mentioning videotex, the *TRAC* decision said the FCC has the discretion, as part of the public interest mandate, to apply the fairness doctrine to new technologies.

Therefore, videotex transmitted over cable television wires may be subject to broadcast content controls. In congressional debate over the fairness doctrine codification bill it was made quite clear that cable television is subject to content controls. Nevertheless, the courts have given cable more First Amendment protection than broadcasting, and a challenge to the FCC's authority to regulate the content of cable-delivered videotex would probably be successful. Furthermore, given the positions of Judges Bork and Scalia in the *TRAC* case, there is a good chance the fairness doctrine itself may be declared unconstitutional in the future. If the courts declare the fairness doctrine to be unconstitutional for broadcasting, it will also be unconstitutional for cable, teletext, and videotex.

(4) Videotex transmitted over telephone wires is not subject to broadcast content controls. The content of videotex messages carried by common carriers cannot be regulated, with the possible exception of obscenity and indecency.

Recommendations

(1) Teletext and videotex should receive the same full First Amendment protections as the print media. They are essentially no different than electronically delivered newspapers. Government content controls should not be applied to teletext and videotex because the controls are unnecessary. Moreover, content regulation would only serve to inhibit the development of these technologies, which hold much promise for consumers.

(2) The fairness doctrine and equal opportunities rules should be repealed. Only then will there be a truly free marketplace of ideas which allows for rapid development of information systems.

★   ★   ★

As this study went to press, the Federal Communications Commission repealed the fairness doctrine in response to an order by the U.S. Court of Appeals in the *Meredith* case. The decision of the FCC has been appealed in the courts, and legislation is anticipated which would again attempt to codify the doctrine.

End Notes

[1]*Comments of the Videotex Industry Association on the Report and Recommendations of the United States Concerning the Line of Business Restrictions Imposed on the Bell Operating Companies, United States v. Western Electric Company,* Civil Action No. 82-0192, at 8 (D.D.C. 1987) (Comments filed March 13, 1987).
[2]P. DeWitt, "Punching Up Wine and Foie Gras," *TIME*, December 1, 1986, at 65.
[3]For an up-to-date directory of companies providing videotex services, *see Public Access Videotex* (Newspaper Advertising Bureau, 1987).
[4]*Id.*
[5]*See* G. Arlen, "High Rollers with High Hopes," *Channels Field Guide '87*, at 86-87.
[6]*See* G. Nahon, "Videotex in the World at the Beginning of 1987," *Videotex Times*, June 1987, at 3.
[7]*Id.*
[8]"Taft Airs Affordable Teletext," *Broadcast Engineering*, Nov. 1983, at 100.
[9]*See* "WBTV's High Hopes For Teletext," *Broadcasting*, Mar. 31, 1986, at 77.
[10]"NBC Launches Teletext Service," *Broadcasting*, May 23, 1983, at 68.

[11] *See* A. Pollack, "Time Inc. Drops Teletext Experiment," *New York Times*, Nov. 22, 1983, at D2.

[12] C. Jackson, H. Shooshan, & J. Wilson, *Newspapers and Videotex: How Free a Press?* 7 (1981).

[13] *See Hoover v. Intercity Radio Co., Inc.*, 286 F. 1003 (D.C. Cir. 1923).

[14] In 1926, a federal court in Illinois held that the Secretary of Commerce had no power to promulgate regulations assigning specific frequencies or limiting the times of operation because this power was not delegated by Congress under the 1912 Act. *United States v. Zenith Radio Corporation*, 12 F.2d 614, 618 (N.D. Ill. 1926).

[15] 44 Stat. 11 Sec. 4 (1927).

[16] *Id.*

[17] *Id.* sec. 18.

[18] 47 U.S.C. sec. 303(r).

[19] *Id.* sec. 303(j).

[20] Communications Act of 1934, 47 U.S.C. sec. 153(b) (Section 3[b] of the Act).

[21] *Id.* at sec. 3(h).

[22] *Freedom of Expression Act of 1983:* Hearings Before the Sen. Comm. on Commerce, Science, and Transp., 98th Cong., 2d Sess. 182 (1984).

[23] 67 Cong. Rec. 5,480 (1926).

[24] *Chicago Federation of Labor v. F.R.C.*, 3 F.R.C. 36 (1929).

[25] *See KFKB Broadcasting Ass'n v. F.R.C.*, 47 F.2d 670, 671 (D.C. Cir. 1931).

[26] 319 U.S. 190 (1943).

[27] *Id.* at 215-16.

[28] *Id.*

[29] 13 F.C.C. 1246, 1256-57 (1949).

[30] *Id.* at 1246.

[31] *Id.*

[32] *Id.*

[33] 395 U.S. 367 (1969).

[34] *Id.* at 390. It has since been learned that Fred Cook was in the employ of the Democratic National Committee and part of a plan to intimidate broadcasters from carrying right wing commentary supportive of Senator Barry Goldwater's presidential campaign.

[35] *See Meredith Corp. v. FCC*, No. 85-1723, slip op. (D.C. Cir. Jan. 16, 1987); *Radio-Television News Directors Ass'n v. FCC*, No. 85-1691, slip op. (D.C. Cir. Jan. 16, 1987).

[36] 104 S. Ct. 3106, 3116 n.11 (1984).

[37] 801 f.2d 501, 517 (D.C. Cir. 1986) *reh'g en banc denied*, 806 F.2d 1115, *pet. for cert. filed*, 55 U.S.L.W. 2200 (U.S. Feb. 20, 1987).

[38] *See Inquiry into Section 73.1910 of the Commission's Rules and Regulations Concerning the General Fairness Doctrine Obligations of Broadcast Licensees*, 102 F.C.C.2d 143, 225 (1985) [hereinafter *Fairness Report*]. On Aug. 4, 1987, in *Meredith v. FCC*, the FCC repealed the fairness doctrine. *See Broadcasting*, Aug. 10, 1987, pp. 39A-K.

[39] *Fairness Report*, 102 F.C.C.2d at 219.

[40] *See* 133 Cong. Rec. S5232 (daily ed. April 21, 1987).

[41] 18 U.S.C. Sec. 1464 (1976).

[42] 413 U.S. 15, 23 (1973).

[43] 438 U.S. 726 (1978).

[44] *Id.* at 748.

[45] Public Notice, FCC 87-153 (April 29, 1987).

[46] *Id.* at 3.

[47] *Id.* at 5.

[48] *See United States v. Midwest Video Corp.*, 406 U.S. 649 (1972); *United States v. Southwestern Cable Co.*, 392 U.S. 157 (1968).

[49] 533 F.2d 601, 608 (D.C. Cir. 1976).

[50] *See First Report and Order* in Docket No. 18397, 20 F.C.C.2d 201 (1969). The fairness doctrine does not currently apply to cable access channels, but would if the FCC ruled such access channels constituted "origination cablecasting".

[51] *See Jones v. Wilkinson*, 800 F.2d 989 (10th Cir. 1986). This decision was summarily affirmed by the Supreme Court.

[52] *Id.*

[53] *See Notice of Inquiry* in Gen. Docket No. 83-989, 48 Fed. Reg. 43,348 (Sept. 23, 1983); *Memorandum Opinion and Order* in Gen. Docket No. 83-989, adopted March 7, 1984, 56 Rad. Reg. 2d (P&F) 49 (1984), *recon. denied*, 56 Rad. Reg. 2d (P&F) 934 (1984).

[54] 129 Cong. Rec. H10,560 (daily ed. November 18, 1983).

[55] No. 85-1160, slip op. (D.C. Cir. Sept. 19, 1986).

[56] The fairness doctrine and equal opportunities rule are applicable to cable television. *See* Cable Television Bureau, FCC, *Cable Television and the Political Broadcasting Laws: The 1980 Election Experience and Proposals for Change* 9-36 (1981). The FCC's statutory authority, however, is unclear. *See* G. Shapiro, *Cablespeech* 65-70 (1983). So long as these rules are applicable to cable, decisions authorizing continued FCC regulation in this area are relevant to teletext and videotex.

[57] *See Radio-TV News Directors Ass'n v. FCC, appeal docketed*, No. 85-1691 (D.C. Cir. Sept. 30, 1986) (appealing FCC's decision to continue enforcing the fairness doctrine despite its recent findings that the doctrine is constitutionally suspect and violates the public interest); *Meredith Corp. v. FCC, appeal docketed*, No. 85-1723 (D.C. Cir. Sept. 30, 1986) (appealing FCC ruling that a Syracuse television station violated the fairness doctrine by running a series of issue advertisements).

[58] *See* 46 Fed. Reg. 60,851.

[59] *Id.*

[60] *Id.* at 60,852.

[61] 53 Rad. Reg. 2d (P&F) 1309 (1983).

[62] The FCC stated: [T]eletext will not be required to further or promote a station's performance with respect to its public service obligations as it relates to programming."

[63] The FCC cited sections 151 and 303(g) of the Communications Act as support for this last proposition.

[64] *Memorandum Opinion and Order*, 101 F.C.C.2d 827 (1985).

[65] 422 U.S. 205, 210 n.6 (1975).

[66] 47 U.S.C. sec. 4(d)(2)(A).

[67] The *current* FCC has been deregulatory in its orientation. All four members have been appointed by President Reagan. An FCC run by a Nicholas Johnson or Charles Ferris would likely be much more regulatory in its outlook and actions. It would likely use every legal tool at its disposal to oversee communication under its jurisdiction.

[68] 102 F.C.C.2d at 169.

[69] *See* "House Turns Investigative Eyes on Network News," *Broadcasting*, May 4, 1987, at 31. Congressman Ed Markey, Chairman of the House Telecommunications Subcommittee, held three hearings, setting the stage for "a serious congressional push to revive the public trust approach of broadcasting regulation."

KF 4774 .B65 1987

	DATE DUE	